CONTAINER decorator

Boxes, baskets and pails
– 25 easy transformations

CONTAINER

decorator

Boxes, baskets, pots and pails
– 25 easy transformations

STEWART AND SALLY WALTON

PHOTOGRAPHY BY GRAHAM RAE
STYLING BY CATHERINE TULLY

LORENZ BOOKS
NEW YORK • LONDON • SYDNEY • BATH

This edition published in 1996 by Lorenz Books an imprint of Anness Publishing Limited
Administrative Office: 27 West 29th Street
New York, NY 1001

Lorenz Books are available for bulk purchase for sales promotion and for premium use. For details write or call the manager of special sales, 27 West 20th Street, New York, NY10011; (212) 807 6739

Produced by Anness Publishing Limited
1 Boundary Row
London SE1 8HP

ISBN 1 85967 118 7

Publisher: Joanna Lorenz
Senior editor: Clare Nicholson
Photographer: Graham Rae
Stylist: Catherine Tully
Designer: Caroline Reeves

Printed in Spain

1 3 5 7 9 10 8 6 4 2

CONTENTS

INTRODUCTION

Containers are essentially practical objects which provide a place to store a range of items from flowers to dog biscuits and dirty laundry to cutlery. The ideas in this book take containers one step further – they are not only practical household solutions but also decorative display items around the home. Everybody has a different approach to structuring their living space, with some people thriving on order and others on chaos. The projects in this book provide a rich source of ideas to help you add a variety of colors, textures, patterns, and shapes to whatever style of organization you opt for. Whichever projects you choose to make, remember to adapt them to your personal taste and use the suggestions in this book as a starting point for your own creativity. Adapt the ideas so that they work best for you in your individual situation. Never underestimate the amount of pleasure and satisfaction that can be gained by taking the creative approach. An afternoon spent drilling, sticking and painting is totally absorbing, and at the end you have a sense of achievement that money just can't buy.

Shopping is fine, but creating is even more fun!

QUICK IDEAS

Leading a frantically busy life should not stop you from enjoying some hands-on homemaking. You don't always need a lot of time for this, and, once you have given it a go, you are bound to find that it is worth the effort. Creativity is as much about your state of mind as actually sitting down and making something – an original idea is a creation in itself. The best way to add originality to your home is not to decorate it in one go, but rather to let it evolve. If you pass a second-hand shop and can spare five minutes, then go inside and see what you can find. You never know – you might just be in the right place at the right time to find something wonderful to start you off on a project.

If you return from the supermarket with interesting foreign packaging, keep the fancy tins and boxes to refill and display on your kitchen shelves. Italian packaging is often decorated with ornate engraved lettering, so buy the best olives or biscotti and, after eating them, you can keep on enjoying their container in another guise.

A quick coat of enamel turns a baked bean can into a bright cachepot for an herb plant on your window sill. A row of these in different colors looks great with an assortment of leaf shapes.

Keep your mind and eyes open when it comes to choosing containers and don't be surprised if you find yourself choosing a plant pot for a fruit bowl or a baby basket for your long johns!

Above: These wire baskets used to be handed out in the changing rooms of English swimming pools and can still be found in second-hand shops today. Put shoes, socks and underwear in them and hang skirts or pants on the crossbar.

Left: This elegant wirework plant container would make a charming fruit bowl for a table centerpiece.

Above: This wire mesh container is an interesting variation on a traditional hat box. The addition of ribbon trimmings makes it ideal for holding bathroom bits and bobs.

Above: It is often the details that make the difference between practicality and style – this wire pail is a wonderful object on its own, but the addition of a layer of straw for the eggs to rest on has made it evocative as well as practical.

Right: There are some wonderful rustic combinations of bark, sticks and thongs about, like this one which has been filled with lemons.

TEXTURES

Above: There are two main types of texture – natural and manufactured. Here, the contrast between the two is exaggerated to dramatic effect with the manufactured wire mesh sieve tray setting off the smooth green apples to perfection.

Above: Old earthenware pots are often overlooked in favor of pottery from more exotic locations. These pots come from the New Forest, England, and were used to transport cider to the fields for the workers.

We respond to texture as we do to sound – we need contrast and variety to satisfy our senses. Imagine a totally smooth or rough environment – both would be strangely monotonous. Containers are an ideal way of introducing different textures into a room, because they are movable and adaptable objects.

You can combine many kinds of natural textures to create a mood. Mix rope with willow baskets, or wooden boxes and burlap bags for a homespun look. Bamboo trays, banana leaf bowls and coconut bark baskets create an oriental mood, while shells, driftwood and pebbles evoke the seashore. This doesn't mean that you have to theme textures, but it is worth noting their mood-enhancing powers. You can also play visual tricks with texture – hard can appear soft if you cover hard wood with a woollen blanket, or you can turn light wood into cold heavy metal with a rusty paint effect.

If you like the cool high-tech look, seek out metal boxes, buckets and canisters. They can always be given a face lift with aluminum spray paint or be bashed around a bit if you prefer a distressed finish. Try tapping a raised pattern into tin with a hammer and nail – ethnic or folk-art motifs work especially well.

Origami boxes can be made from any paper that is thin enough to fold and crease. There are some wonderful natural fibre papers around that come in rich earthy colors and their rough textures provide another dimension to the crisply folded shapes. Experiment with metallic or high gloss paper for a completely different look.

Whatever containers you decide to make or adapt, always think about texture in the same way as you would color or shape. It can make all the difference.

Above: Basketware from the Far East is often made from twigs and sticks. The left-hand basket could be used as a wastepaper basket, and the right-hand one could be used as a container for holding plant pots or fruit and vegetables.

Left: Wire mesh wastepaper baskets used to be found under every desk and they can still be bought in second-hand shops today. Wire has made a big comeback, and an original basket or bin will have far more character than a new one.

Above: The simple lines of origami boxes make the most of the natural texture of the paper.

EASY EMBELLISHMENTS

Above: A diversity of textures can look stylish, especially when the materials are natural. The coarse rope handles provide a contrast against the polished surface of the wooden box. An electric drill was used to make four sets of small holes in the sides, then the handles were tied in place with twine, knotted inside the box. Try fixing leather handles to a basket for a similar contrast of natural textures.

One way of making containers look a bit more special is to add decorative features to an existing box, basket or bucket. The shapes of these utilitarian household objects have been pared down to a good basic design and they provide the starting point for some dressing up and enhancing.

The essential piece of equipment for doing this type of work is a glue gun. Once the heat has liquified the glue stick, it can be dotted on to almost any surface to provide an instant bond. A glue gun makes a huge difference to the sort of decoration you can apply to a vertical surface. A clay flowerpot, for example, can be decorated with shells or moldings in a matter of minutes – the glue sets as it cools so there is no need to work in sections and wait for each to bond before turning the pot. It is important to plan your design arrangement and mark all positions in advance because the glue sets in seconds.

Instead of buying decorated containers for your home, start looking for plain ones. This way, all the hard work has been done for you, and all that's left is the fun of applying the finishing touches.

Left: Ribbons are not just for presents. They are strong enough to stand up to regular use, and they come in so many wonderful patterns and colors that choosing them can be a real problem. Use ribbons as fastenings, decorations, and trimmings, and enjoy the sensual pleasure of tying and unraveling beautiful bows.

Above: These balls make an interesting alternative to a standard handle.

Left: Plasterwork moldings are available from interior decorating shops. They can be painted, distressed or gilded and then added to a container to give it a period feel.

Left: Natural decorations like shells, starfish and driftwood can turn an old box into a seashore treasure chest. Experiment with different arrangements, then use the glue gun to stick them in place. This easy and instant technique can be used with anything from wooden balls to fake fruit.

PAINT EFFECTS

Above: This box was painted a rich, deep blue. The sun face was then traced onto the top using transfer paper and colored in a metallic gold felt-tipped pen. When the design was finished, a few coats of clear varnish made the colors look even richer.

A simple coat of paint gives an instant lift to an old can or box, but if you use a special paint effect the object is transformed. Color, texture and pattern can all be bought over the counter nowadays, and there is plenty of fun to be had in their creative application.

You can trace lettering, numbers or patterns on to cans or boxes using transfer paper. It has one chalk-impregnated side that goes on the surface of the container, then you simply place the tracing on top and re-draw it.

Antiquing varnish comes in different shades, but it is basically used to make things look older and more yellow. A freshly painted container can be rubbed with steel wool or sandpaper in the places where most wear and tear would naturally occur – around the edges, near catches, handles or on the lid. A coat of tinted antiquing varnish on top will complete the illusion.

Special paint effect kits contain everything you need to produce the effect and they can be bought in small sizes suitable for container projects. Many shops now sell sets of unpainted wooden boxes known as "blanks" that are ideal for this sort of treatment. There are also mass-produced molded plastic flower boxes and urns that can be given a verdigris, old terracotta or rust finish.

Paint effects can make wood look like gold and tin look and feel like lizard skin when you crackle glaze it. Everything is not what it seems, and that's what makes it such fun.

Left: This is a genuine antique example of a bargeware painting. The colors have faded with time, but each brushstroke that makes up the flowers is still visible. To give an authentic and soft look to your painting, mix subdued shades and use the bright ones as highlights. Rub the finished painting with steel wool and coat it with a tinted varnish to simulate age.

Above: You can transform kitchen containers by applying a basecoat of metallic paint in a hammered metal finish. Mix a spoonful of silver powder into clear satin varnish and sponge it on to the containers to add a sparkle to the shine.

Left: Instead of burying these peat pots in the ground for seed cultivation, give them a paint treatment. Undercoat them with terracotta latex paint, then lightly dry brush them with a lighter color, so that the terracotta still shows through.

Above: This sewing box was decorated in a traditional Shaker style, where the first coat of color, a maize yellow, is covered by a dragged coat of brick red. The red was rubbed back in places with fine sandpaper to simulate wear and tear. The drawer front has a coat of tinted varnish combed into a squiggly bird's-eye maple pattern. A small piece of dowel and a screw serve as the handle.

ADAPTING CONTAINERS

Above: Oil cans have a very distinctive bird-like shape. They make unusual vases, most suited to single strong dramatic shapes like a thistle or pink gerberas. One oil can may look accidental on its own, but two together are a definite statement.

The idea for adapting a container can be as simple as using a bottle for a vase or displaying your logs in a tin bath. It is more about taking a sideways look at things than actually making anything new.

A really crazy example of an adaptation was once sighted in a small front garden – it was a car being used as an oversized flower trough. The roof had been removed and flowers spilled out, cascading down the sides and over the hood! In fact any number of different items, such as chimney pots, watering cans and baskets, can be used as containers for plants. You may start off by using your hat to contain the fruit you've picked, then realize how good it looks on the kitchen table and keep using it as a fruit bowl. A garden sieve is great for freshly dug onions, carrots and potatoes, and three new sieves chained together make an excellent hanging vegetable rack.

Oranges and lemons are sold in bright plastic mesh bags that can be pegged on a rail in a child's room to hold, but not hide, building bricks, cars and puzzles. String bags can be used in the same way to organize seasonal clothing, like gloves, hats and scarves – not quite out of sight, but certainly out of the way.

Beach buckets in the bathroom, oil cans full of flowers, chocolate boxes packed with sewing equipment and goldfish bowls stuffed with silk scarves – when it comes to adapting containers, absolutely anything goes!

Left: This top hat has been made from dyed bark. The combination of colors is intense and the bark feels like strong fibrous paper. As this is not an everyday hat or a put-away one either, it is used to display a collection of bright South American juggling balls – a great combination.

Right: It is really simple to transform old cans into excellent containers for candles. Richly decorated Mediterranean cans are ideal.

Below right: These brightly painted catering size tins make bright and cheerful containers for plants.

Below: This vivid lime-green basket has an unusual shape and is both dazzling and practical. The plastic-coated wire was looped and bent into a shape that flatters everything it contains. Eggplants and lemons form a potent mixture of textures and colors here, but imagine it also filled with red tomatoes or orange clementines.

MEDITERRANEAN CRATES

Wooden fruit and vegetable crates are much too good to be thrown away once emptied, so rescue them and dress them up with color and ribbon to make a great set of containers. Take a trip to your local vegetable store or wholesale vegetable market and pick the best crates available. The crates here work well, as they have a solid base that can be separated and used as a lid. These rustic Mediterranean crates look wonderful stacked with candles or tablecloths.

YOU WILL NEED

- **3 wooden fruit or vegetable crates**
- **sandpaper**
- **pliers and wire cutters**
- **staple gun (optional)**
- **powder paint: red, blue, and green**
- **paintbrushes**
- **40in checked ribbon**
- **scissors**

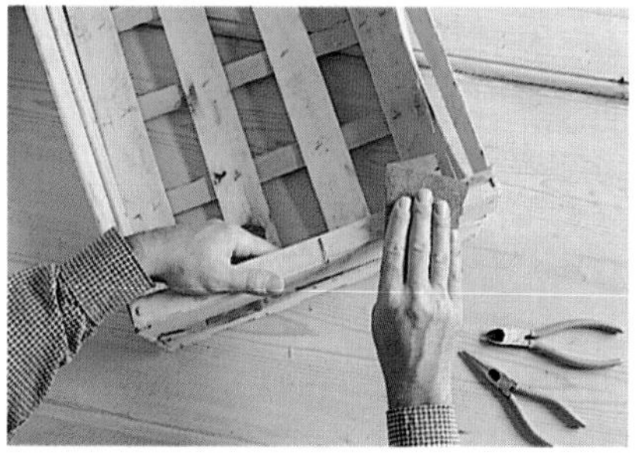

1 Rub off any rough edges on the crates with sandpaper. Detach the base from one of them to be the "lid". Remove and replace any protruding staples if this is necessary.

2 Mix the powder paints according to the manufacturer's instructions. Paint one of the crates red inside and blue outside, and green along the top edges.

3 Paint the other crate green inside and blue outside, and red along the top edges.

4 Paint the lid blue. Bind the six center joints with crosses of checked ribbon, tied at the back.

TERRACOTTA LAUNDRY POT

Laundry in a flowerpot? It sounds unusual, but this idea makes a refreshing change from wicker baskets in the bathroom. Terracotta pots are now available in a huge range of shapes and sizes and a visit to your local garden center should provide you with just the right pot. To give a pristine pot an antique feel, follow these simple steps.

YOU WILL NEED

- ♦ large terracotta flowerpot
- ♦ rag
- ♦ shellac button polish
- ♦ white latex paint
- ♦ paintbrush
- ♦ scouring pad
- ♦ sandpaper (optional)

1 Soak a rag in button polish and rub the surface of the pot with it. The polish will sink in very fast, leaving a yellow sheen.

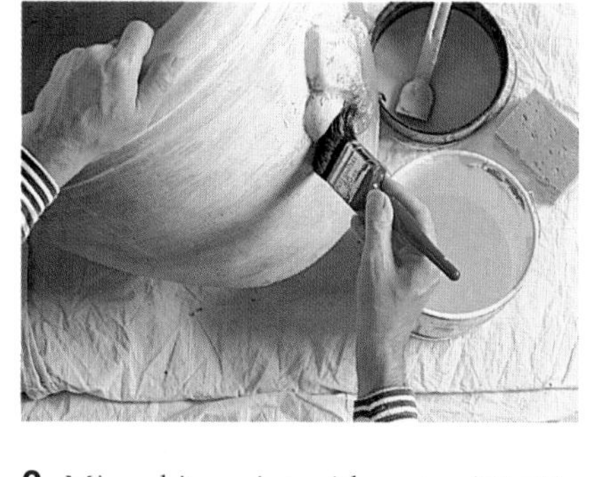

2 Mix white paint with water (50:50). Stir it well, and apply a coat to the pot. Allow to dry.

3 Rub the pot with the scouring pad to remove most of the white paint. The paint will cling to the crevices and along the moldings to look like limestain. Either leave the pot like this or rub it down further with sandpaper to reveal the clay.

4 When you are happy with the effect, apply a coat of button polish with a brush to seal the surface.

WINDOW BOX SHELF

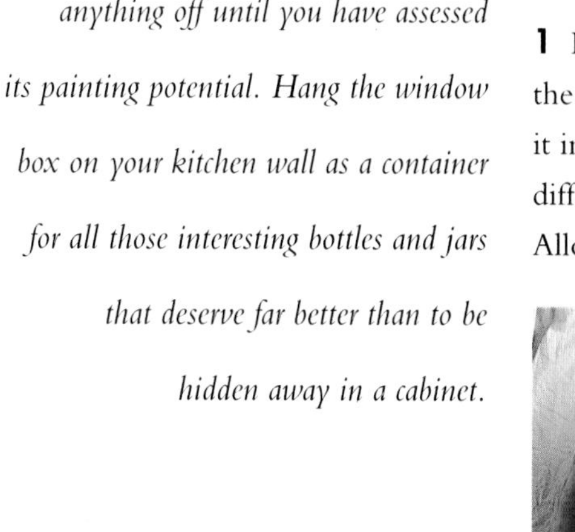

This rough-hewn rustic box was discovered, stained dark brown, in a junk shop, and was probably hanging in a garden shed about thirty years ago. It serves as a good reminder not to write anything off until you have assessed its painting potential. Hang the window box on your kitchen wall as a container for all those interesting bottles and jars that deserve far better than to be hidden away in a cabinet.

YOU WILL NEED

- **wooden window box**
- **sandpaper**
- **latex paint: red, blue, green, and white**
- **paintbrushes**
- **fine-grade steel wool**
- **shellac button polish**
- **drill, with wood bit**
- **2 wallplugs and screws**
- **screwdriver**

1 Rub down the wooden surface of the window box with sandpaper. Paint it in bright colors, highlighting different parts in contrasting colors. Allow to dry.

2 Mix white paint with water (50:50) and apply a coat of this all over the window box. Allow to dry.

3 Rub down the white paint with steel wool so that it just clings to the wood grain and imperfections.

4 Apply a coat of button polish to protect the surface from stains and to improve the aged effect. Drill two holes in the back of the window box and fix it to the kitchen wall.

Packed in England
expressly for Williams-Sonoma Inc.
San Francisco, CA. 94120.

HANGING JAM JARS

This bright idea for getting double mileage out of kitchen shelves is borrowed from an old-fashioned tool shed. Woodworkers and gardeners used to line the undersides of their shelves with jam jar lids and then screw in jars filled with nuts, bolts, washers, springs, and other such useful items. Everything was on view and within easy reach. Fill your jars with cookies, candies or different types of beans and lentils. You could even use smaller jars to create an herb-and-spice rack.

YOU WILL NEED

- **4 or more jam jars, with lids**
- **kitchen shelf**
- **pencil**
- **awl**
- **screws (no longer than the shelf depth)**
- **screwdriver**

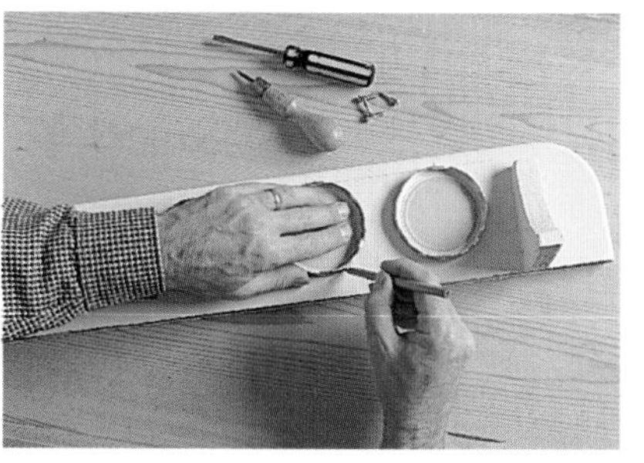

1 Arrange the jam jar lids along the underside of the shelf. There should be sufficient room between them for a hand to fit. Lightly mark the positions with a pencil.

2 Use an awl to make two holes through each lid into the shelf.

3 Screw the lids in place.

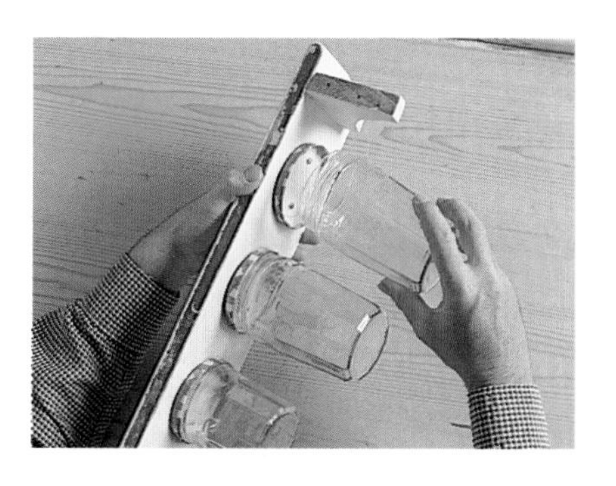

4 Screw on the jars. It is easier to do this with the shelf on a work surface than with the shelf already hanging on the wall. Hang up the shelf.

CAMEMBERT NUMBERS

This project should be linked to a wine and cheese party or the befriending of a French restaurateur as it involves nine Camembert boxes. It is always a relief to be able to recycle packaging and these empty boxes are certainly given a new lease on life in the form of a stylish set of containers. Enlarge the numbers on a photocopier so they fit your boxes and cut them out as stencils. If you have never cut a stencil before, it is easier to cut stencil cardboard than plastic, which tends to be slippery. The craft knife needs to be very sharp, so keep fingers out of the way while you cut. If you find the task too daunting, trace the numbers onto the boxes and paint them in.

YOU WILL NEED

- **9 Camembert boxes**
- **fine-grade sandpaper**
- **clear matte acrylic varnish**
- **paintbrushes: medium, stenciling, fine-pointed, and square-tipped**
- **number templates, see page 93**
- **craft knife**
- **cutting board**
- **spray adhesive**
- **stenciling plastic or waxed stencil cardboard**
- **enamel paint: 9 colors and silver**
- **pair of compasses and pencil (optional)**
- **shellac button polish**

1 Remove any paper labels from the boxes. Rub the wood with sandpaper to get rid of any rough edges.

2 Apply a coat of clear matte acrylic varnish to all the boxes.

3 Take the templates or, to make your own, enlarge the templates at the back of the book on a photocopier so the numbers are 3in high (or to fit within a border of about ⅝in). Cut each one out on a cutting board.

4 Use spray adhesive to stick the photocopies onto the stencil material. If you are using plastic, stick them underneath and if you are using cardboard, stick them on the top.

5 Cut out the stencils on the cutting board. Make the incisions away from the corners and turn the stencil rather than trying to cut at a different angle.

6 Stencil a number in the middle of each lid using a different enamel color each time.

7 Use the nine colors and a fine-pointed brush to paint a band around the top of each lid. If necessary, use a pair of compasses to draw the band about ⅝in wide. Choose colors that contrast well with the numbers.

8 Using a square-tipped brush, carefully paint a band of silver around the lid edges.

9 Allow the boxes to dry, then rub them down lightly with sandpaper.

10 Apply a coat of button polish as a final seal. This gives the boxes a yellowish tinge and a matte sheen.

Right: If you wanted to make these boxes for children to store things in, paint plus, minus and equals signs on extra boxes so that they can practice their math. Equally useful would be boxes with letters of the alphabet painted on.

THREE-TIER VEGETABLE RACK

Most kitchen accessories seem to be either made from practical but unattractive plastic, or high-tech and expensive chrome or stainless steel. This handsome black vegetable rack is not the product of a design team from Tokyo, Milan or Paris – although it looks as if it could be – but is in fact three garden sieves (made either from wood and wire or galvanized iron) and a length of gatepost chain! Vegetables are natural show-offs which look wonderful on display in a shop, but once stored in a cabinet or fridge drawer at home are inclined to turn nasty. Suspend this rack in a kitchen corner so you can reach vegetables easily for cooking – be warned, though, they might look too good to disturb for the sake of a meal!

YOU WILL NEED

- 3yd chain, plus length to hang it from ceiling
- hacksaw and pliers
- string
- 3 metal garden sieves
- felt-tipped marker pen
- scissors
- masking tape
- cloth
- center punch
- hammer
- drill, with metal and wood bits
- 15 "S" hooks (optional)
- eye and ring
- ceiling hook and fixings (or long wall bracket)

1 Saw the chain into nine sets of twelve links only sawing through one half of each link.

2 Pry the links apart with pliers to separate the nine sections.

3 Wrap a piece of string around one of the sieves to measure the circumference. Mark the measurement on the string. Cut the string to length.

4 Fold the string into three equal lengths, marking it with the pen as you do so.

5 Place the marked string around one of the sieves and stick masking tape on each third around the edge. The other two sieves need six marks – three round the top edge and three around the base edge. It is important to do this accurately so that the rack hangs level.

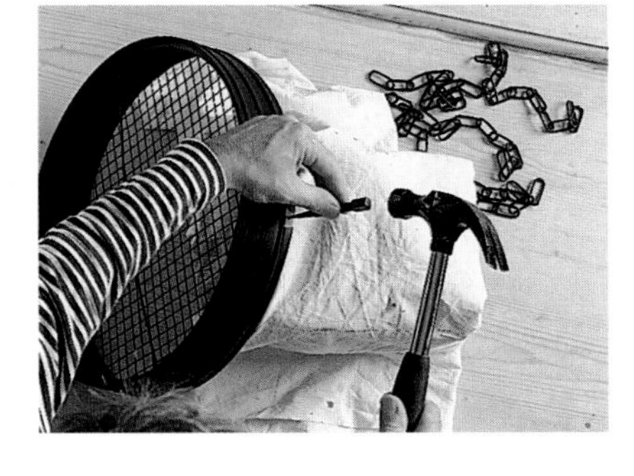

6 Place the sieves on a hard surface covered by a cloth. Use a center punch with a hammer to dent the metal at each marked point.

7 Using the dent marks, drill holes in all three sieves. The dent mark should prevent the drill bit from slipping on the metal.

8 Insert "S" hooks into all the holes. Arrange the sieves in the order in which they will hang – the three-holed sieve goes at the bottom. Starting with the three-holed sieve, attach a length of chain to each "S" hook, adjusting the level to suit your needs by attaching the hooks at different links. To save on money, you could dispense with the "S" hooks and simply pry apart and then re-join the chain links to attach the sieves.

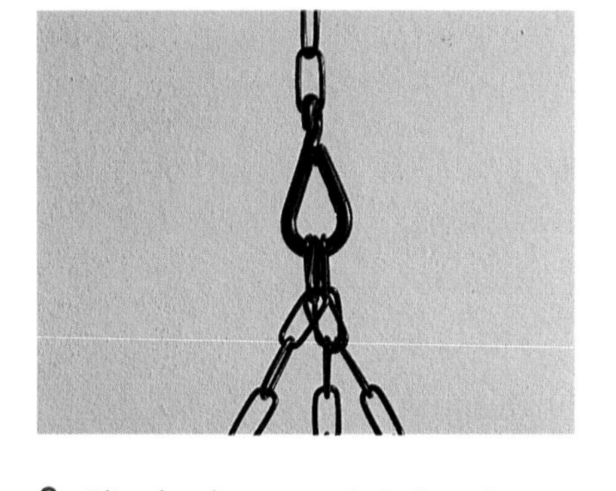

Right: Multi-tiered racks are the perfect container if you are short of space, and they can have a function beyond the kitchen. Many apartments have small bathrooms with no ledge around the tub to keep shampoos, soaps and sponges. These racks are perfect for hanging over the tub and holding all these odds and ends.

9 Clip the three top chain lengths together with an eye and ring, and add a further length of chain to suspend the rack from the ceiling. Find a beam for the ceiling hook as it needs to be good and strong.

Left: The vegetable rack will look equally at home in a stark, modern interior.

BATHROOM BUCKETS

Add an element of seashore fun to your bathroom by hanging up a row of bright buckets. This trio of enamel-painted buckets was bought from a toy shop, but you could take a trip to the seashore where you are bound to find a great selection of buckets in all shapes and sizes. While there, go for a stroll along the shore to find the ideal bit of driftwood to fix them to your wall.

YOU WILL NEED

- 3 enamel-painted buckets (or plastic beach ones)
- length of driftwood (or an old plank)
- pencil
- drill, with wood and masonry bits
- wire
- pliers
- wire cutters
- masking tape
- wallplugs and screws
- screwdriver

1 Line the buckets up at equal distances along the wood. Make two marks, one at each end of the handle where it dips, for the three buckets.

2 Using the wood bit, drill through the six marked positions to make holes through the wood.

3 Wind wire around each handle end and poke it through the holes. Twist the two ends together at the back to secure. Trim the ends. Then drill a hole near each end of the wood.

4 Hold the piece of wood in place and mark the positions for the fixings. Place a small piece of masking tape over the tile to prevent it from cracking, then drill the holes and fix the wood to the wall.

BABY BASKETS

These generous-sized baskets were designed to transport babies in great comfort. Sadly, they do not conform to any safety regulations and present-day newborns have safe plastic and metal contraptions instead. Baby baskets are used here as fresh and airy containers for clothes, such as socks and underwear.

YOU WILL NEED

- plank of wood
- tape measure
- saw
- sandpaper
- drill, with wood and masonry bits
- 1 long or 2 short branches (or poles)
- penknife (or wood carving knife)
- dowel (optional)
- 2 baby baskets
- wood glue
- hammer
- wallplugs and screws
- screwdriver

1 Cut two squares of wood at least 4¾ x 4¾in and 2in deep. Sandpaper the edges and drill a hole through the middle of each square, slightly smaller than the diameter of the branches. Carve away the branch ends so that they fit tightly into the holes. Sandpaper it slightly.

2 Use offcuts from the branches or pieces of dowel to make the pegs. Taper the ends.

3 Measure the distance between the basket handles and drill two holes that far apart on top of each branch. The holes should fit the pegs correctly. Apply wood glue to each branch and tap them into each square, making sure that the pegs are on top. Apply wood glue to the peg ends and fit them into each branch.

4 To fix the top branch to the wall, drill holes in the four corners of the square of wood and four corresponding holes in the wall. Using the wallplugs and screws, screw the branch to the wall. Repeat to fix the lower branch to the wall, allowing about 4in clearance from the top basket.

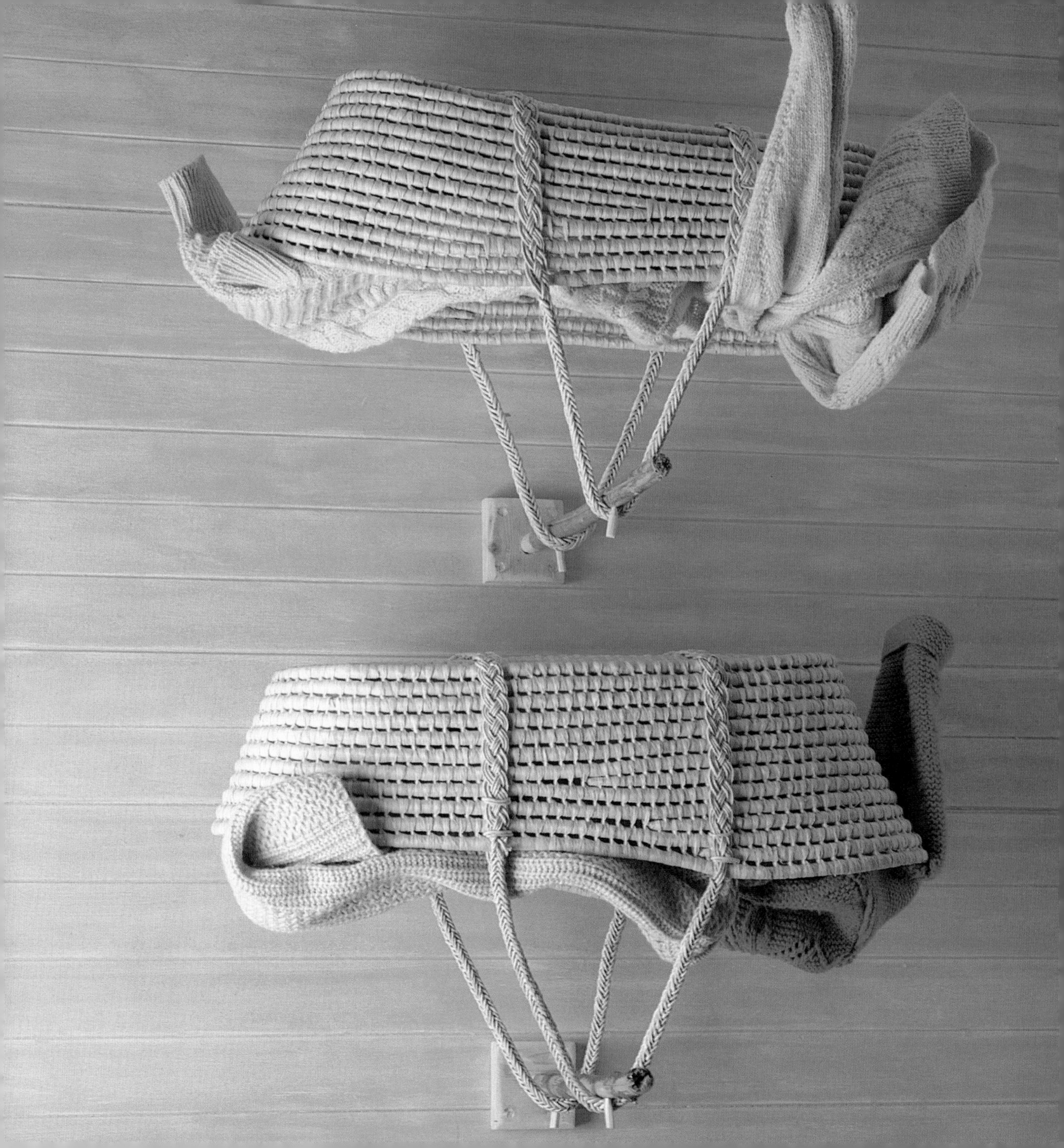

SCULPTED JUG

This project pays homage to that great, and often underestimated, material – chicken wire. Sculptors have used chicken wire on the insides of their creations for many years because of its ability to take on and hold shapes. They cover the wire with bandages or cheesecloth before building up the surface with clay and eventually casting it in metal. Now, chicken wire is used as a sculptural material in its own right. You can adjust the size and shape of your creation as often as you like to suit your needs. If you want to use the wire jug for flowers, simply put a glass inside it to hold the water.

YOU WILL NEED

- **roll or piece of chicken wire**
- **tape measure**
- **wire cutters**
- **protective gloves (optional)**
- **pliers**
- **bottle**
- **newspaper**
- **enamel car spray paint: 2 or 3 colors**
- **scrap cardboard**

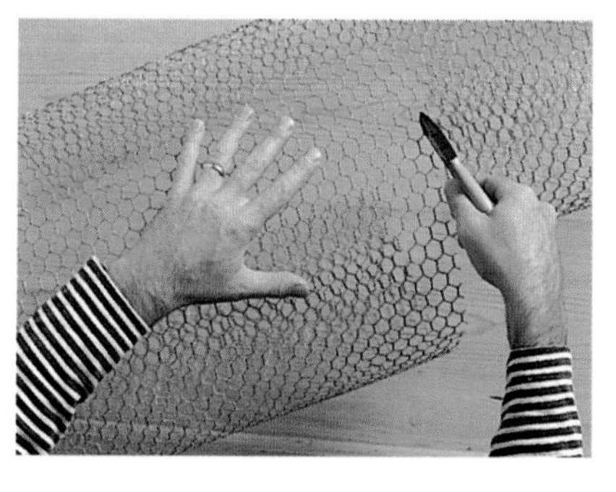

1 Using wire cutters, cut a piece of chicken wire 40 x 20in. The edges of chicken wire are very scratchy so it is advisable to wear protective gloves.

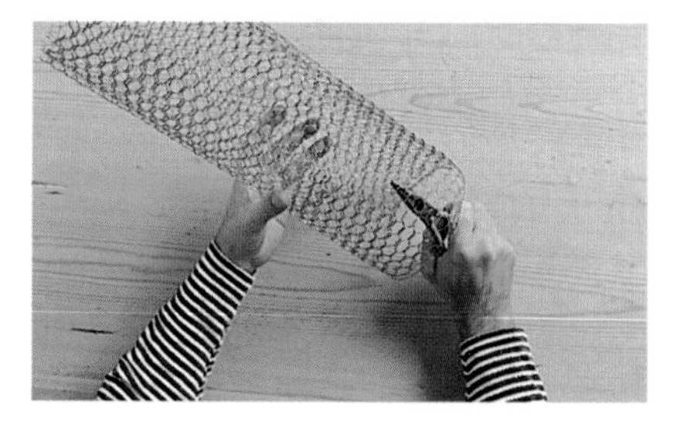

2 Roll the wire into a double-thickness tube, starting from the long edges. It will stand 20in tall.

3 For the base, squeeze the sides of the individual wire "modules" together. Start by slightly squashing every one in the fourth row from the bottom so the base begins to taper inward. Gently curve this row, but for the next and bottom two rows squeeze more modules together to make the base flat. Use a bottle to push the base into shape from the inside.

4 Cut the outer tube at the top into six strips that reach halfway down, then cut the inside tube into six strips that go a third of the way down.

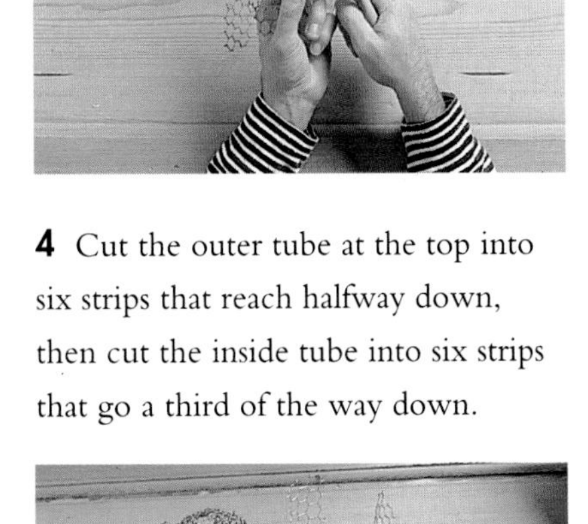

5 Bend the outer strips down all around the tube.

6 Squeeze the modules together at the point where the outer strips begin, to pinch in a good "waist." Leave a bulge. Then pinch the tube in again where the inner strips begin.

7 Fold in the sharp edges of the lower strips, and pinch them down firmly with the pliers.

8 Curl these strips inward until they rest against the wide part of the chicken wire jug.

9 Curl up the top strips, so that they curve outward and their ends meet the tube to form the top of the jug.

10 Cut a separate piece of wire measuring 4¾ x 12in, to make the handle and roll in the sharp edges.

11 Shape the handle into a curve.

12 To attach the handle, use small bits of wire and twist them between the body of the jug and the base and top of the handle. Tuck the ends in and squeeze them together.

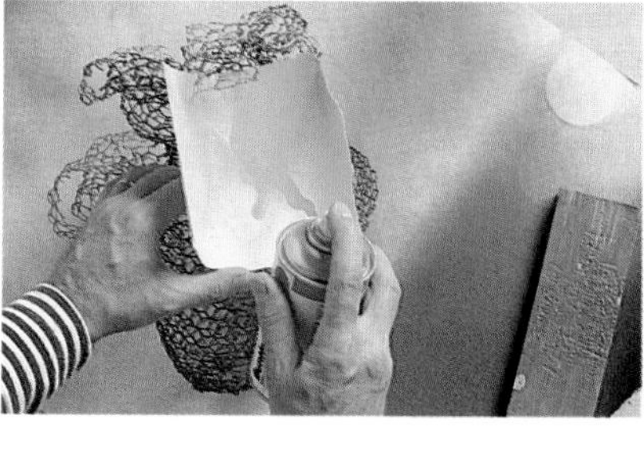

13 Place the jug on newspaper in a well-ventilated area. Spray it with paint, masking areas off with cardboard to get the effect you require.

Right: Chicken wire is a fantastic medium as it is cheap, attractive and readily available. Its malleability and lightness make it suitable for creating large structures.

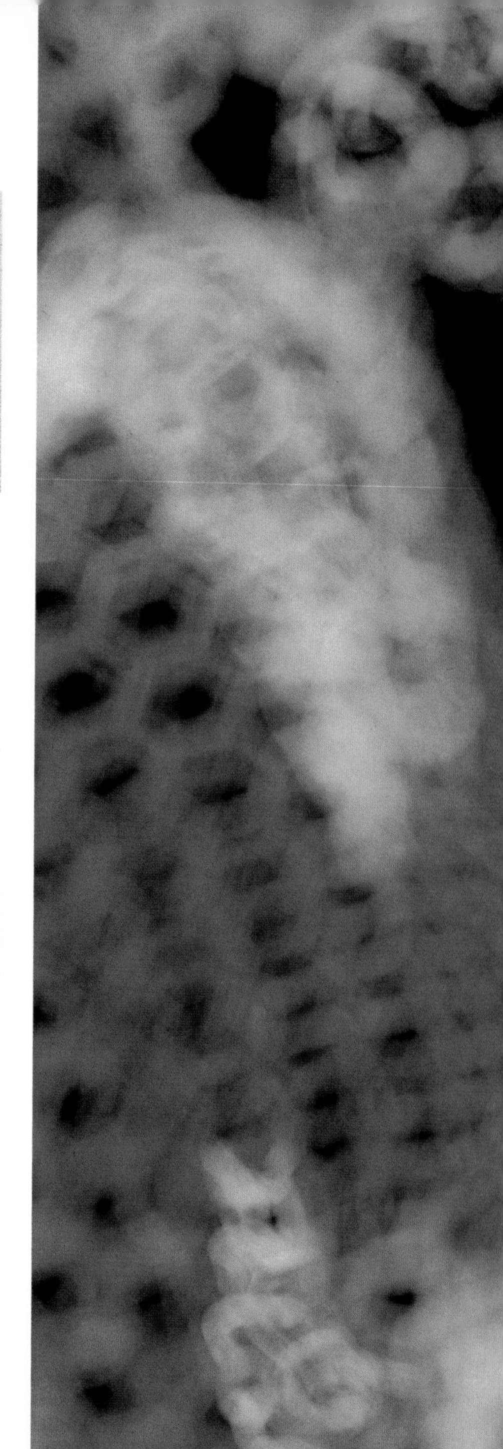

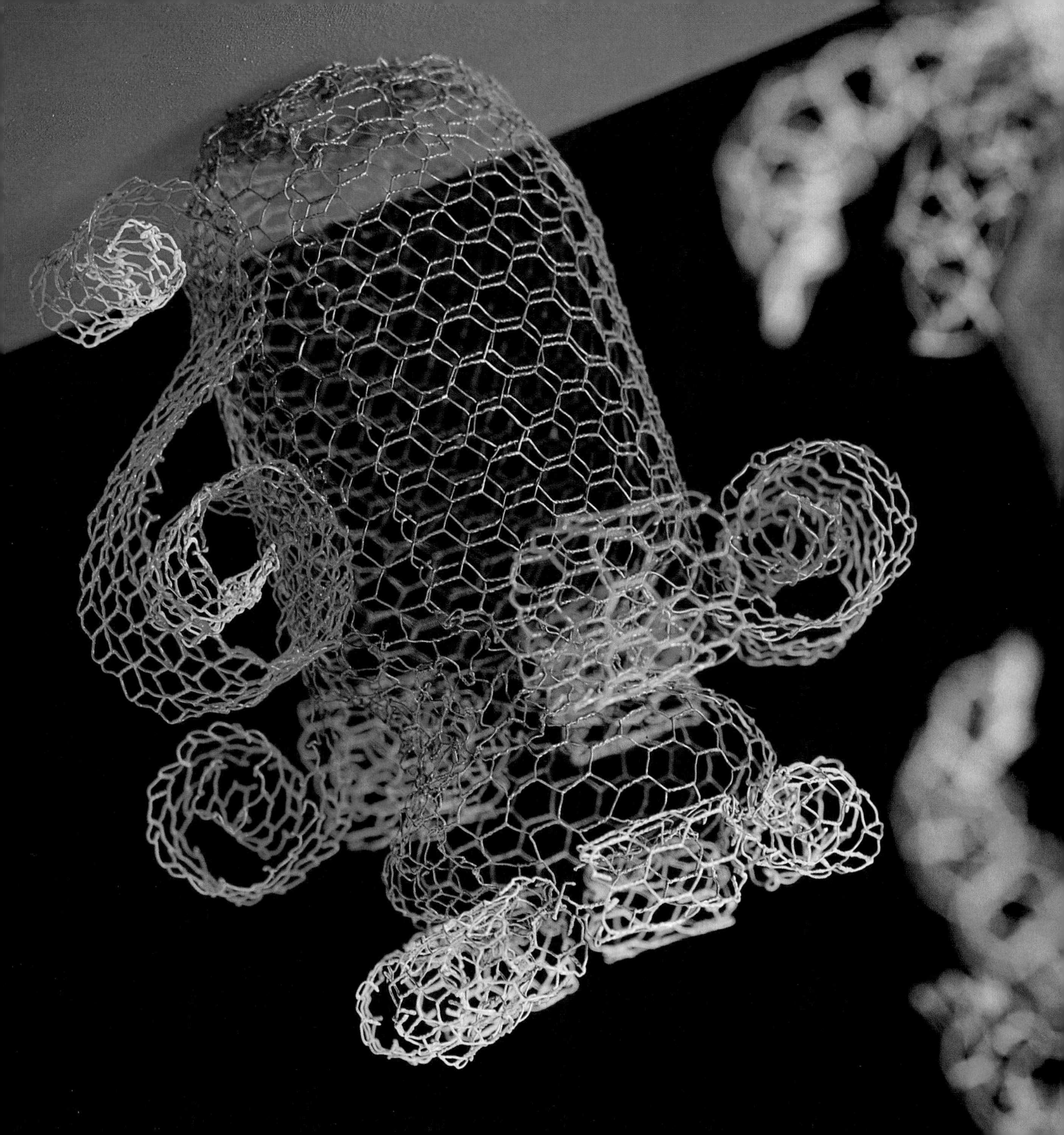

DUSTPAN DESK ORGANIZER

If you have to excavate piles of paper every time you need a pen, then it's time to get the dustpans out! Two metal dustpans can be spray painted in any color combination and fixed together with bolts. Use the dustpans for stationery, a small in/out tray or, most suitably, for bills!

YOU WILL NEED

- **2 metal dustpans**
- **4 metal washers**
- **enamel car spray paint: lime green and metallic blue**
- **newspaper**
- **G-clamp**
- **2 pieces of corrugated cardboard**
- **pencil**
- **drill, with metal bit**
- **2 nuts and bolts**
- **pliers**
- **screwdriver**

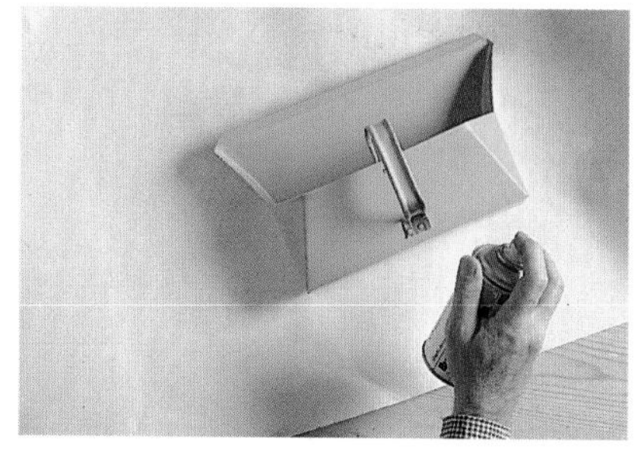

1 Spray one dustpan and two washers lime green and the others metallic blue. Protect your work surface with newspaper and build up the color gradually with light puffs of paint. Allow to dry.

2 Clamp the dustpans together, protecting the paintwork with cardboard. Hold the washers where the dustpans meet and mark two equal positions with a pencil.

3 Drill through the marked pencil positions to make two holes through both dustpans.

4 Place a washer on each side of one hole, hold the nut with pliers to prevent it from scratching the paintwork, and tighten the bolt with a screwdriver. Repeat for the other hole.

ORIGAMI BOXES

Origami is the Japanese art of paper-folding, where the most extraordinary three-dimensional objects are made from a single folded square of paper. In theory, once you understand the basics, you can make anything from an elephant to a vase of flowers without ever resorting to glue or tape to hold it together. This little box is a suitable project for a complete beginner, but it may take a couple of practice attempts before it suddenly "clicks." You can make boxes any size you like, and use any paper that is not too flimsy. Package wrap is good as it creases well and is both cheap and strong, so practice on that to begin with. Once you have perfected the technique, you can move on to colored papers and make a range of containers.

YOU WILL NEED

- selection of paper squares for practice (optional)
- sheet of paper
- scissors
- small piece of double-sided tape (this breaks all the rules!)

1 Fold and trim a sheet of paper to make a perfect square.

2 Fold the square corner to corner, and open it out again so you can see the lines of four equal triangles folded on the paper.

3 Turn the paper over then fold this into four square quarters, then unfold again so you can see eight equal triangles folded on the paper. The center of the paper is Point A.

4 Hold the model in the air and push all the sides together so that the corners meet in the middle.

5 Flatten the model. Point A is now a corner of the flattened model. Keep it facing you.

6 Fold the point opposite Point A over to meet it. Carefully crease it along the mid-line.

7 Now fold it back up to the mid-line and crease the fold. Unfold it again.

8 Fold the same point up to the last crease. Then fold it over again, up to the mid-line. Turn the model over and repeat steps six to eight.

9 With point A still facing you, take the top layer of the left-hand corner and fold it over on to the right-hand corner. You are now faced with a square. Fold the left-hand corner and the top layer of the right-hand corner into the center point, and crease. Bring the right-hand top layer over on to the left. Turn over and repeat.

10 Point A now forms the base of a smaller triangle. Fold this up at the point where it meets the sides. Crease well. Then fold it in the other direction and crease again. Unfold.

11 To form the box shape, insert your hand through the open end. Splay out your fingers while pushing up the bottom of the model, Point A. The box shape appears with two folded edge sides and two taller pointed sides.

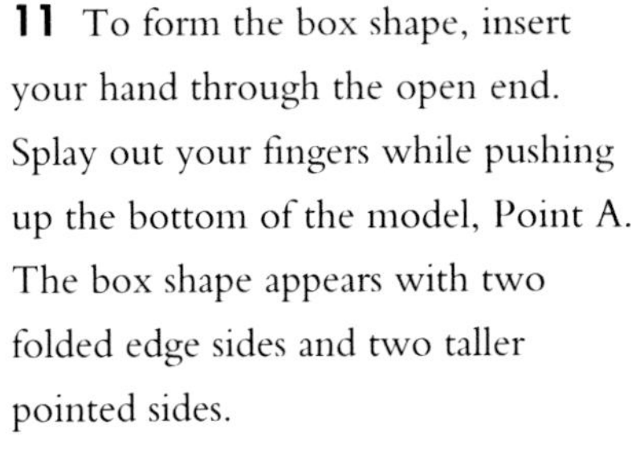

12 Fold the two pointed sides down to be level with the others and tuck the ends under the box. These ends can be secured with a small piece of double-sided tape.

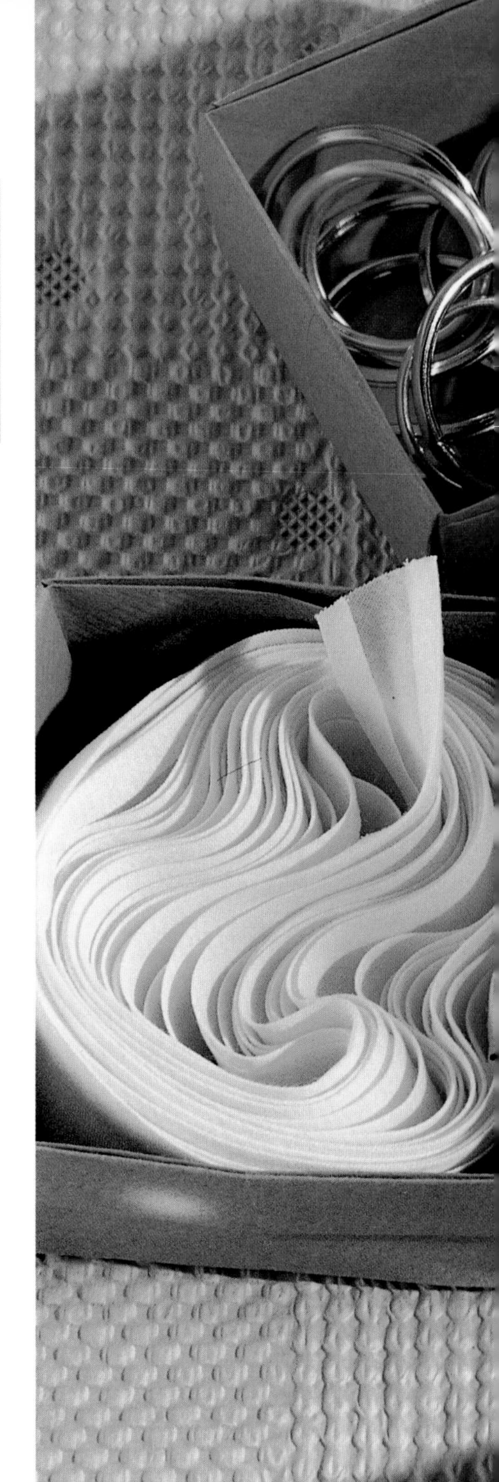

Right: This box was adapted from an origami model with a handle. The changes made necessitated the use of double-sided tape which is not usually permitted in the art of origami.

STRING BOTTLES

Liqueur bottles have such lovely shapes that it seems a shame to put them into the recycling bin. This method of recycling enables you to keep on enjoying the bottles even after you have enjoyed their contents! The bottles used here are sherry, crème de menthe and Armagnac. This project is very easy to do and it can almost seem like therapy once you've gathered together a ball of string, some glue, a pair of scissors and three interestingly shaped bottles. All that you have to do then is make yourself comfortable, put on some relaxing music and start winding the string around the bottles.

YOU WILL NEED

- ball of string
- glue gun, with all-purpose glue sticks (or all-purpose glue)
- 3 interestingly shaped bottles
- scissors

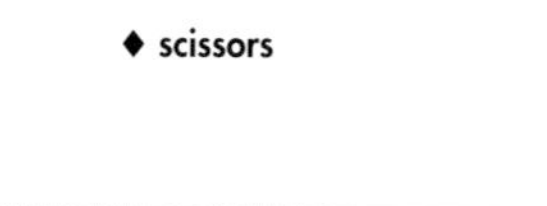

1 Coil one end of the string around like a coaster. Place a dot of glue in the center of the bottle base. Heat the glue gun and apply glue in spokes over the base. Press the string onto them. Draw a ring of glue around the edge to make the base secure.

2 Circle the bottle with the string, working your way up and applying glue as you go. Make sure you get a good bonding on the bends.

3 When you reach the top of the bottle, cut the end of the string and apply plenty of glue to it so the finish is neat with no fraying. Repeat these steps with the other bottles.

SHOES IN NEWSPRINT

Special shoes deserve a home of their own and these sturdy wooden wine boxes can be made stylish enough to house anything from Oxfords to glass slippers. The boxes were lined with different types of newsprint, which was layered and stuck down like papier mâché. Select the newsprint to suit your shoes – put leather lace-ups in the pink financial pages, party shoes in comic strips and velvet pumps in the arts and literary review pages.

YOU WILL NEED

- **wooden wine crates**
- **sandpaper**
- **white glue**
- **paintbrushes**
- **variety of newsprint**
- **craft knife**
- **clear matte varnish (or shellac button polish)**

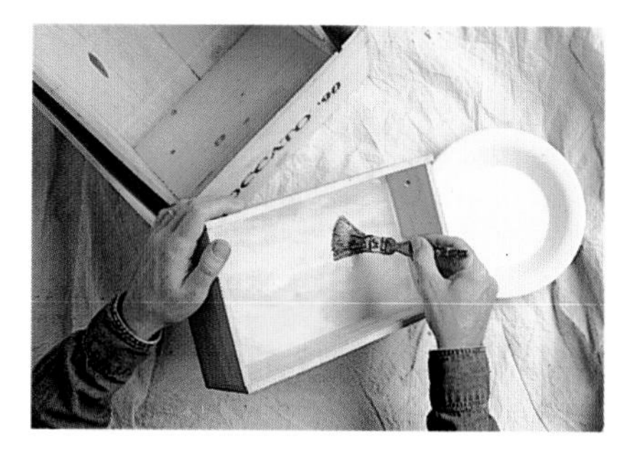

1 Sand down a wine crate. Then mix white glue with water (50:50) and apply a coat to the inside.

2 Apply another coat of the glue mixture. Then smooth newsprint all over the inside. The glue will be absorbed by the paper. Apply undiluted glue along the top edges and smooth the paper over it. Allow to dry before applying more paper and glue to make a random all-over pattern.

3 Leave until bone dry, then trim the paper along the outside top edges with a craft knife.

4 Varnish the whole box with either clear matte varnish or, if you want an "aged" look, button polish. Leave to dry, then recoat at least twice.

BEADED LAUNDRY BASKET

Brightly colored plastic laundry baskets are cheap and practical, but they need help to give them a more individual look. This purple laundry basket was made glamorous with bright Chinese checker pieces taken from an inexpensive children's set. They are ideal, as the domed pieces have spiked backs that can be trimmed to the required depth and glued into holes drilled in the plastic. The fixing is very secure, so don't worry about beads scattering all over the bathroom floor!

YOU WILL NEED

- Chinese checker pieces
- wire cutters
- plastic laundry basket with lid
- drill, with fine bit
- glue gun, with all-purpose glue sticks (or all-purpose glue)
- masking tape

1 Sort the Chinese checker pieces into colors. Trim the spikes so the pieces will fit into the depth of the plastic basket without protruding on the other side.

2 Guided by the shape of the basket, drill holes for the spikes in a circle around the lid and in stripes down the sides.

3 Sort the pieces into the color sequences you want to use. Heat the glue gun, apply a dab of glue to a spike and push it immediately into a drilled hole. The glue will set right away, so work fairly quickly as you fill all the holes.

4 Run a length of masking tape around the base of the basket as a positioning guide for a straight band of colored pieces. Fix these as you did the other pieces.

DOG BISCUIT BOX

Boost your pooch's self-esteem with this tough container for his snacks. It is the canine equivalent of a padlocked cookie jar and is guaranteed to stop any late- night raids! The "stone" is a kit of plaster-cast slabs that is used to decorate fireplaces. Although the manufacturers recommend a special cement, you can use grouting instead.

YOU WILL NEED

- wooden box with lid
- gray latex paint
- paintbrush
- dark gray coursed stone-cladding kit
- pencil
- coping saw
- glue gun, with all-purpose glue sticks (or all-purpose glue)
- tile grouting
- black or grey powder paint
- tool for applying grouting

1 Paint the inside of the box gray, then arrange the slabs around it. Mark any overlaps that need to be sawn off.

2 Cut the slabs down to size with a hacksaw. Plaster cuts easily but it does make a lot of dust. Wear a mask for protection.

3 Heat the glue gun and stick the slabs in position. The glue will set right away, so work quickly. Once set, saw off any pieces that will prevent the box from sitting evenly on the floor or the lid from fitting well.

4 Mix up the grouting, adding powder paint to tint it dark gray. Fill in the gaps between the slabs.

BLANKET CHEST

You can almost guarantee that every interesting pine chest has been discovered by now, stripped and sold for a profit, but there are still plain, solid work chests around that can be used as a good base for this project. The blanket used for covering the chest is the utilitarian sort used by furniture removal firms as a protective wrapping in their vans. Any blanket would be suitable, but this sort has lots of "give" because of the way it is woven and so can be stretched for a smooth fit. The chest has a piece of upholstery foam on it so it doubles up as a comfortable bedroom seat. The lid is held down by a leather strap – suitcase straps or old horse tack are ideal as they come in longer lengths than leather belts.

YOU WILL NEED

- wooden chest
- screwdriver
- pliers
- tape measure
- blanket
- dressmaker's scissors
- staple gun
- upholstery foam, to fit lid
- ruler
- cutting board
- craft knife
- upholstery tacks
- scrap card
- small hammer
- small piece of leather (or cardboard)
- leather strap, to fit around chest

1 Unscrew and remove the hinges from the lid of the chest. Use pliers to remove any protruding nails or screws.

2 The chest is lined and covered with a single piece of blanket. Measure around the chest for the length of the blanket. Then measure the height of the chest. Double the height measurement and add 5in to allow for tucking the blanket into the inside and under the base.

3 Cut the blanket to your measurements. Spread out the cut blanket flat on the floor, find the middle and lie the chest on its side on the blanket. There should be an even amount of blanket each side of the chest and 3in of blanket protruding out from underneath the base.

4 Cut from the front edge of the blanket, in a straight line, to the left and right front corners of the chest. Fold the cut section inside the chest and staple it.

5 Smooth the blanket down the side and under the base. Staple it under the base with a line of adjoining staples.

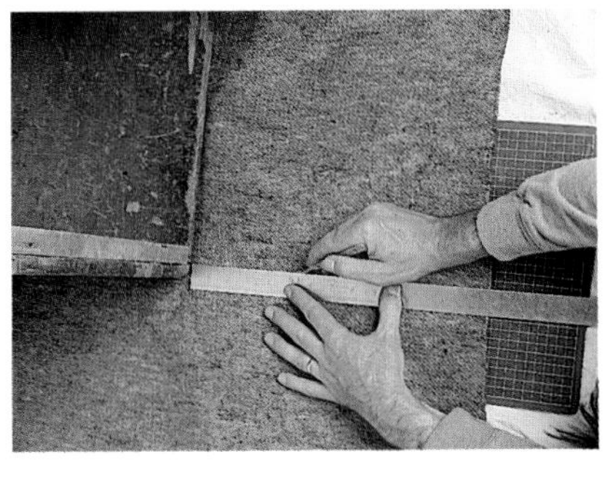

6 Turn the chest on one end and cut the blanket from the edge down to meet both corners, as you did for the front. Smooth it inside the chest and staple it. Do the same on the outside of the end and under the base. Repeat this on the other end.

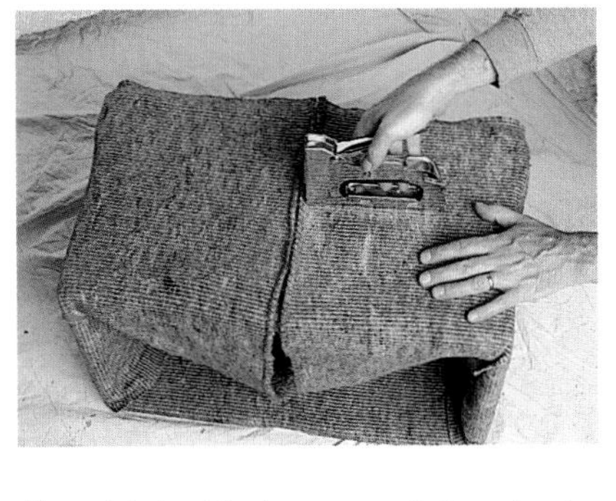

7 Fold the blanket around from both sides so that the edges meet at the back. Staple down the back, then fold the lining inside and staple it.

8 Staple all the lining neatly down inside the chest. A good way of finishing this off is to cut an extra piece of blanket to fit the inside base and staple it around the edges.

9 Cut a piece of blanket about 4in larger than the chest lid on all sides. Place the piece of upholstery foam in the middle of the blanket with the wooden lid on top. Press down on the foam and pull up the blanket on one side. Then put in a few staples to hold it in place.

10 Cut a triangular section off each corner, leaving enough blanket between the lid corner and the cut to fold up and staple onto the wooden lid of the chest.

11 Fold up the joined edge while pushing down on the corner to compact the foam. Staple it across the corner. Then pull the side pieces up and over the first piece. Staple them in place.

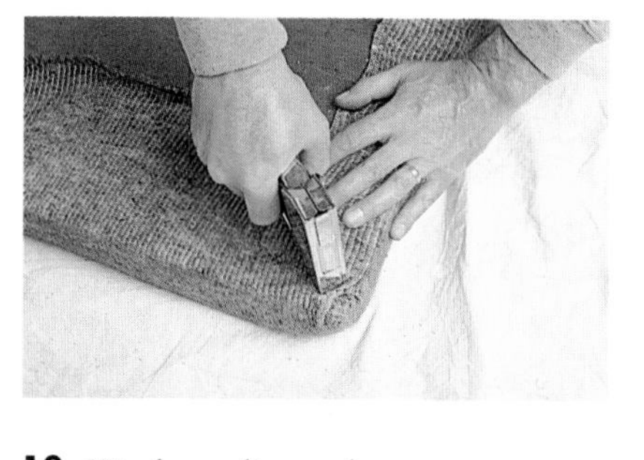

12 Work on diagonal corners alternately because the blanket stretches and you need to get the tension right. Neaten the corners and edges by folding and trimming. As with the base, you can cut another piece of blanket just smaller than the lid and staple it over to cover the edges.

13 Decide upon the spacing of the upholstery tacks, then cut a cardboard strip to use as a guide between them. Hammer in the tacks, protecting their heads by placing a piece of leather or cardboard over them as you do so.

14 Use the tacks to highlight the shape of the chest and to secure the fastening strap to the lower half.

Right: Transformed beyond recognition, this box now looks stylish and functional.

LIZARD-SKIN BOXES

These smart angular boxes look crisp and exclusive, but they are in fact no more than paper-covered foam coreboard! They make great containers for jewelry, headbands or cufflinks and they would add a touch of elegance to any desk or dressing table. When you design your own boxes, there are no manufacturing constraints, so you can make them any shape you like. This is a freedom to enjoy, so forget about the usual squares, circles, and rectangles, and make a random-shaped box unlike any other. Foam core board is light and easy to cut and stick. Buy it from art and craft suppliers, and buy lizard-skin paper from specialty stationery stores.

YOU WILL NEED

- **foam coreboard**
- **cutting board**
- **felt-tipped marker pen**
- **ruler**
- **craft knife**
- **glue gun, with all-purpose glue sticks (or all-purpose glue)**
- **3 different lizard-skin papers**
- **spray adhesive**

1 Place the foam core board on a cutting board and mark out a four-sided angular shape for the base. Cut out the shape.

2 Measure each of the sides of your shape, then decide on the height of your box. Cut out four rectangular side sections to fit those measurements.

3 Heat the glue gun and run a thin strip of glue along each base edge, then stick on the sides. The glue will set right away, so set to work quickly and accurately.

4 Place the box on a sheet of foam core board and draw around the shape. Add 1¼–1½in all round for the lid overlap. Cut out this shape.

5 Do the same again, but this time take off ½in all around. This is the shape for the lower half of the lid to fit inside the box.

6 Glue the two lid sections together.

7 Cut a strip of lizard-skin paper wide enough to line the inside and outside of the box and to fold underneath it.

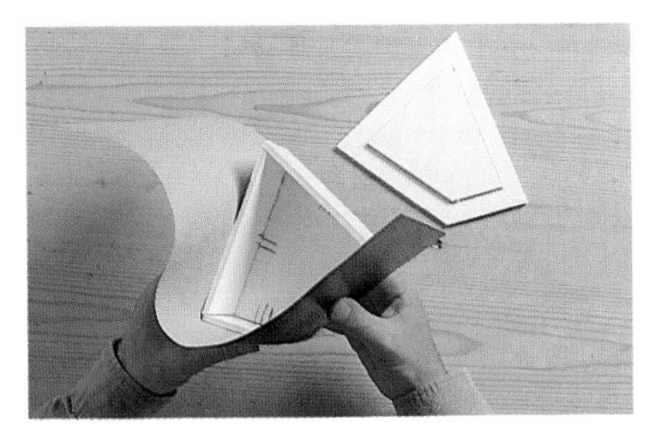

8 Apply a thin coat of spray adhesive to the lizard-skin paper. Then wrap the paper around the box.

9 Cut down into the corners and fold the paper inside the box. Do the same underneath, smoothing the paper flat onto the box. Apply a bit more glue with the glue gun where necessary.

10 Place the lid over the paper, and cut out a shape with a sufficient overlap to cover all the white board on the underside. →

11 Glue the paper onto the lid and cut the corners to make a neat fit.

12 Cut two identical small angular shapes for the handle and glue them together. Cover the handle with lizard-skin paper, and glue it on to the top of the box. Repeat all these steps to make the other boxes.

Right: Lizard-skin paper comes in a range of lovely colors so choose papers that will go with the color-scheme of your home.

BUTTON BOX

Button boxes are an old-fashioned delight that should not be allowed to disappear altogether. There was a time when most homes had a cookie tin filled with an assortment of buttons for sewing and knitting projects. You can make your own button box to store these little treasures. Buttons are wonderfully tactile things to handle and children will spend hours sorting the shapes, colors and sizes of the contents of your button box.

The wooden box used here, with a sliding lid, is an empty tea container, covered with black felt and decorated with buttons. This is a good project to do on a winter's evening as it combines creative pleasure with practicality and is excellent stress therapy as well!

YOU WILL NEED

- wooden box with sliding or hinged lid
- black latex paint
- paintbrush
- black felt
- chalk
- dressmaker's scissors
- rubber-based fabric glue
- craft knife
- cutting board
- buttons
- glue gun, with all-purpose glue sticks (or all-purpose glue)

1 Paint the box black, inside and out. Allow to dry.

2 Place the box on the felt. Use chalk to draw the shapes needed to cover it: one rectangular strip to cover the base and long sides up to the grooves for the lid; one strip to cover the two ends and the base a second time; and finally one strip for the lid that stops short of the runners that fit into the grooves.

3 Cut out the marked felt pieces.

4 Spread fabric glue onto the base and sides of the box and smooth the shorter strip of felt on to it. The felt stretches a bit at this stage.

5 Spread a thin strip of fabric glue along the top edge of the sides and fold the felt over it. Allow to dry.

6 Trim off any edges for a neat finish. Glue the longer strip of felt onto the base and up both of the ends in the same way. Trim off any excess.

7 Cover the lid with felt. Then begin arranging the buttons to make an attractive design.

8 When you are happy with the design, use the glue gun to stick the buttons in place. The glue will set right away, so work quickly and place the buttons accurately as they cannot be moved once stuck.

9 Decorate the sides of the box with stripes of colored buttons.

10 Glue a row of white and pearl buttons along the top edges of the sides to complete the design.

Right: If you have no particular need for collecting buttons, then this idea could be adapted to make a box that reflects a different hobby, such as a stamp-covered box for a stamp collector, a shell-covered box for a shell collector, or a map-covered box for a map enthusiast.

TWO-DIMENSIONAL VASE

This vase looks two-dimensional and is apparently made of paper, but it contains fresh flowers. The trick is to make the vase without a base and place it over a glass bottle or jar. For best effect, the vase should be sited so it can be viewed from either the front or the sides. Select an illustration of any urn, vase or jug that has good lines, and simply enlarge it on a photocopier.

YOU WILL NEED

- **photocopied enlargement of image**
- **spray adhesive**
- **2 A4 sheets of fiberboard (from art and graphics suppliers)**
- **craft knife**
- **cutting board**
- **dark gray stiff paper**
- **felt-tipped marker pen**
- **glue gun, with all-purpose glue sticks (or all-purpose glue)**

1 Apply spray adhesive to the wrong side of the photocopied illustration, then stick it onto a sheet of fiberboard. Carefully cut out the vase. Stick gray paper onto the other sheet of fiberboard.

2 Place the cut out vase on top of the other fiberboard sheet, gray side down. Trace around the shape and cut it out. Cut six equal strips of fiberboard to link the sides together.

3 Dab glue onto the strips and place them in key positions. Start at the sides of the base, standing the vase up to get an even base.

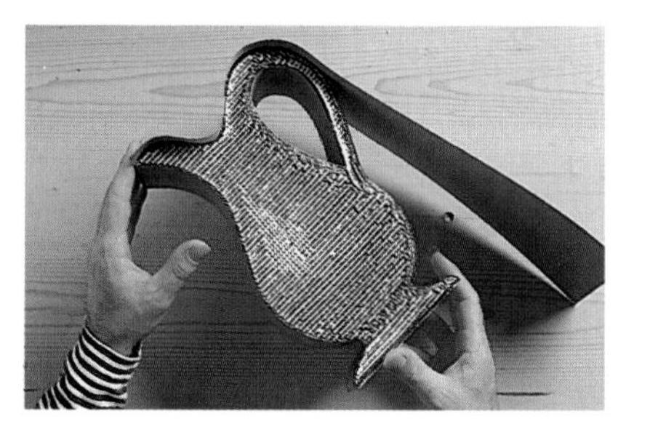

4 Cut a strip of gray paper to cover the longer side from top and base, and another to cover the other side to the base. Place the longer strip over the top and mark the center point. Make the hole for the flowers. Apply glue along the edges of the board and smooth paper strips onto them. Allow to dry before trimming off any excess.

GILDED BOXES

There is no paint or spray that gives a finish to compare with gold leaf, or its cheaper counterpart, Dutch metal leaf. It glows and glitters and can be burnished with a soft cloth to leave a gleaming, reflective surface. The art of gilding is surrounded by mystique because it requires a great deal of skill to lay the fragile sheets of gold leaf onto gold size or glue. This design was created using a kit suitable for beginners, which can be bought from art and craft suppliers. The flat surface of the box makes it remarkably easy to apply, but it is important to follow the manufacturer's instructions as the drying time is crucial to the success of the project.

YOU WILL NEED

- design tracing or photocopy
- spray adhesive
- high-density foam
- craft knife
- round painted wooden box with lid
- fine-grade sandpaper
- pencil
- gold-leaf kit
- paintbrush
- small foam roller
- cotton batting
- fine steel wool (optional)

1 Select a design from a source book, and photocopy or trace to size. Spray the back of the design with adhesive and stick it on to a block of foam.

2 Carefully cut out the copied design, starting in the middle. Then cut out around the outside.

3 Once the cutting is complete, peel off the paper pattern.

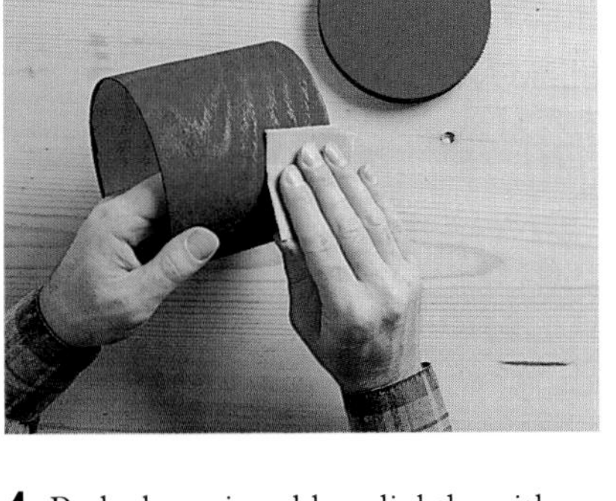

4 Rub the painted box lightly with sandpaper to make a rough surface for applying the size.

5 Draw a steady pencil line around the top edge of the lid. Use your fingers against the box to keep the line an even distance from the edge.

6 Paint a strip of size between the edge of the lid and the pencil line.

7 Put some size onto a plate and run the roller through it until it is evenly coated. Using the roller, coat the foam stamp.

8 Stamp an evenly-spaced pattern around the side of the box, recoating the stamp after each print. Stamp four shapes inside the lid border line and one in the center.

9 Leave the size until it has the right degree of "tack" (according to the manufacturer's instructions) then invert the lid onto a sheet of gold leaf.

10 Smooth the gold leaf over the edges so that it is in contact with all of the size. Apply the sheets of gold leaf around the side of the box in the same way. It is easier if you hold the box at an angle and place the sheets on top.

11 Rub the surface with cotton batting. All the "un-sized" gold will flake off, leaving the design behind.

12 If the brand new gleaming design overwhelms you, rub the box lightly with steel wool to give an aged look.

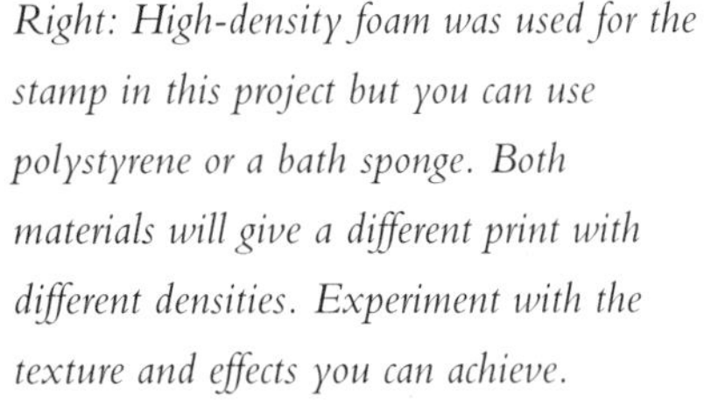

Right: High-density foam was used for the stamp in this project but you can use polystyrene or a bath sponge. Both materials will give a different print with different densities. Experiment with the texture and effects you can achieve.

CIRCULAR PAINTED BOXES

This project experiments with three different paint effects. The first, rust, looks especially good when used on a material that doesn't naturally rust, such as wood. Verdigris is a natural substance that forms on the surface of weathered brass and is a beautiful turquoise-green color. The third finish used here is crackle glaze (craquelure), which looks a bit like lizard skin. This is the most time-consuming of the three paint effects because of the drying time needed between coats, but it is easy to do and the final result emerges excitingly at the end. Try at all three effects and see which turns out to be your favorite!

YOU WILL NEED

- **3 wooden boxes, with lids**
- **water-based paints: dark gray, 2 shades of rust, green-gray, stone, 2 shades of green, and maize yellow (according to required finishes)**
- **handful of fine sand**
- **paintbrushes**
- **3 foam sponges**
- **clear matte varnish, craquelure base varnish, craquelure varnish**
- **alizarin crimson oil paint**
- **2 rags**
- **turpentine**

The rust finish

1 Mix sand into the dark gray base coat to add texture. Apply two coats of paint to the box. Allow to dry.

2 Dab on the darker of the two rust colors using a foam sponge. Cover most, but not all, of the background with a mottled coat of this color.

3 Dab on the lighter rust color. If you are in any doubt about how it should look, find some real rust and copy it.

4 Finally, dab in just a touch of green-gray. Do not overdo this as it should blend in rather than stand out as a sharp contrast.

The verdigris finish

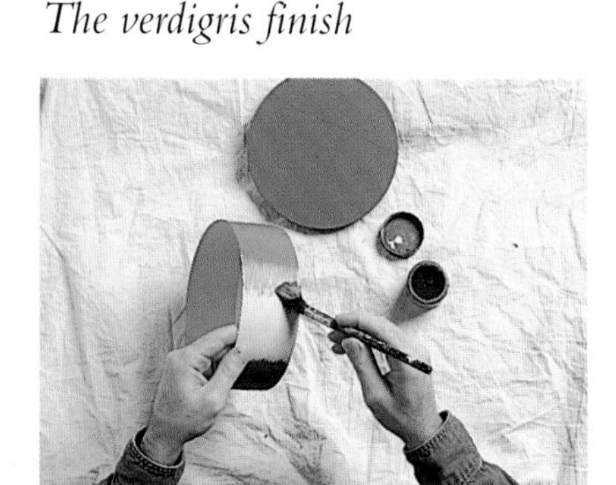

1 Paint the box with the stone base coat color.

2 Dab on the lighter green using a foam sponge. Cover most, but not all, of the background with this.

3 Dab on the other green color using a foam sponge.

4 Apply a coat of clear matte varnish to protect the surface.

The crackle glaze finish

1 Paint the box with the maize yellow base coat color. →

2 Apply an even coat of craquelure base varnish. Brush it out smoothly so that no drips form. Allow to dry for 15–20 minutes.

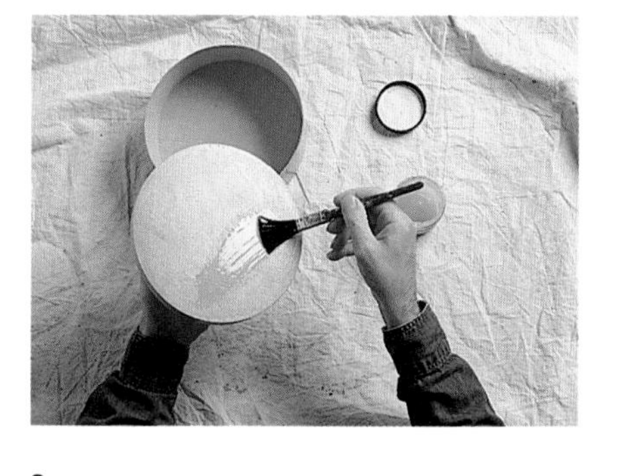

3 Next, use a clean, dry brush to apply an even coat of craquelure varnish. This is the coat that crackles. Allow to dry for 15–20 minutes.

4 Put a small amount of crimson oil paint onto a rag and rub it into the cracks on the surface of the box. Dip a clean rag in turpentine and rub the surface of the box so that the red paint only remains in the cracks.

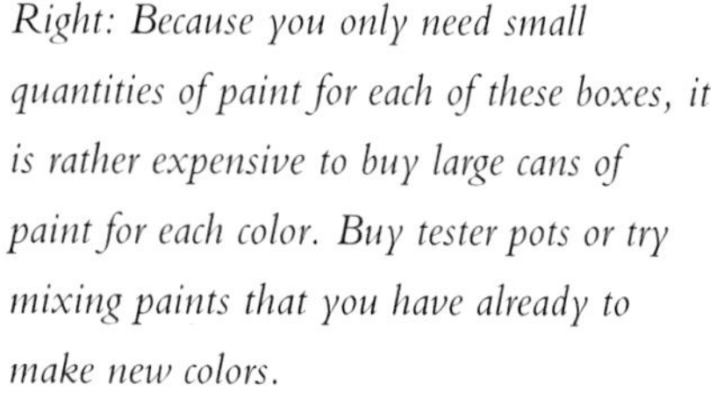

Right: Because you only need small quantities of paint for each of these boxes, it is rather expensive to buy large cans of paint for each color. Buy tester pots or try mixing paints that you have already to make new colors.

SEASHELL BOX

Boxes decorated with seashells can be fairly awful, and unfortunately the worst ones have given this wonderfully relaxing pastime a bad name. But don't be put off by the cheap and commonplace variety – shells are naturally beautiful and there are endless ways of arranging them tastefully. This project combines the contemporary look of corrugated cardboard with a dynamic shell arrangement. For the finishing touch, the box is painted pure white matte. The result is a seashell box that resembles a meringue-topped cake.

YOU WILL NEED

- selection of seashells
- round corrugated cardboard box with lid
- glue gun, with all-purpose glue sticks (or all-purpose glue)
- white acrylic paint (gesso primer is ultra white)
- paintbrush

1 Lay out all the shells and sort them into different shapes and sizes. Arrange them on the lid to make the design.

2 Remove the top layer of shells from the middle of the lid and begin sticking them on. Heat the glue gun, and glue the outside shells first, gradually moving inward.

3 Work with the shell shapes, building up the middle section. The glue gun allows you to get an instant bond, so the shells will stick to the surface anyway you like.

4 Paint the box and the lid white. If you are using acrylic gesso primer, two coats will give a good matte covering; ordinary acrylic or latex paint will benefit from an extra coat.

PUNCHED METAL BUCKET

The idea of decorating metal objects with raised punched patterns has been around ever since sheet metal was invented about 300 years ago. Bare metal buckets are ideal for this sort of pattern-making, and all you need is a pen to draw your guidelines and a hammer and blunt nail for the punching. You can practice your technique on any tin can to find the ideal sort of tap needed to make a good bump without piercing the metal.

YOU WILL NEED

- **bare metal bucket**
- **felt-tipped marker pen (not water-based)**
- **blunt nail (or center punch)**
- **hammer**
- **rag**
- **lighter fluid (or similar solvent)**

1 Draw your pattern on the inside of the bucket. These motifs come from South America, but any repeated curves or angles are suitable.

2 Rest the bucket on the piece of wood to protect your work surface. Tap the nail with the hammer, keeping the dents a regular distance apart. About ½in is fine.

3 Continue hammering the pattern all over the inside of the metal bucket.

4 Use a rag and lighter fluid to clean off the marker pen pattern that is left between the punched marks.

FANCIFUL SHOE BOX

If you are one of those people who always takes the shoe box home with new shoes, only to throw it away reluctantly a while later, then this is the project for you! Your instincts to take the box in the first place are right, as shoe boxes are the perfect shape and size to make useful containers. This box was covered with brown wrapping paper – yet more good recycling – that was rolled and twisted, then unraveled and stuck onto the box to provide an interestingly textured surface. The "bark" is made from torn strips of white paper, coated with wood stain to give a streaky finish. The end result is a unique, natural-looking container suitable for anything from potpourri to an index card system.

YOU WILL NEED

- shoe box
- cream latex paint
- paintbrushes
- brown wrapping paper
- pre-mixed wallpaper paste and cheap brush
- scissors
- white glue
- thick, white paper
- wood stain (such as antique pine)
- thick, coarse string
- piece of unbleached cotton, 4 x 4in
- bulldog clips (or pegs)
- glue gun, with all-purpose glue sticks (or all-purpose glue)

1 Paint the shoe box with cream paint until all of the lettering is covered. Allow to dry.

2 Roll up a length of brown paper, crumpling it as you go. Fold up the roll, then twist it into as small a shape as possible. Carefully untwist and open out the crumpled paper.

3 Apply a coat of wallpaper paste to the box. Place it centrally on the brown paper.

4 Fold the brown paper around the box, pressing it into the pasted surface, but not smoothing it too much. Pinch the paper along the edges of the box and cut along these. Fold the end flaps inside the box, sticking them in place with wallpaper paste.

5 Fold the brown paper around the sides of the box, one end at a time, pasting one on top of the other to create two large triangular shapes.

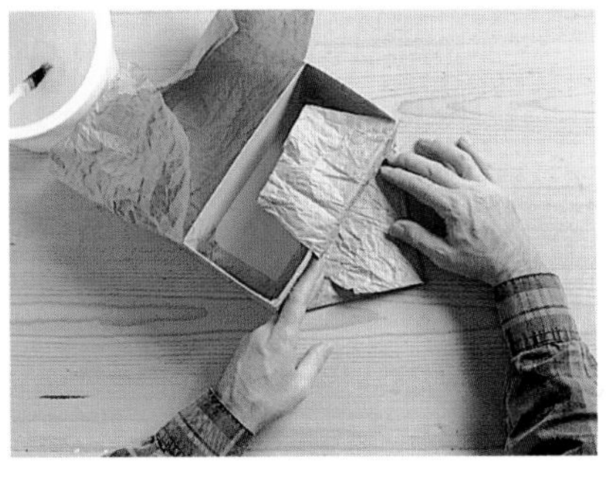

6 Fold the triangular shapes up over the side and paste them against the inside of the box.

7 Neaten the inside by cutting a piece of brown paper to fit the base exactly. Paste it over the paper edges.

8 Using a dry brush, paint a streaky coat of undiluted white glue on the white paper. Leave some of the paper unpainted. Allow to dry.

9 Brush a coat of wood stain onto the white paper. It will be resisted by the white glue where it is at its thickest, and part-resisted in other places. This gives the bark effect.

10 Tear the paper into rough triangular shapes. If you tear at a slight angle, the paper will rip through its thickness and make the edges white and thin. Paint these white edges with wood stain so that they blend in.

11 Roll up the paper triangles, beginning with the widest part and rolling toward the point. Bundle the strips together with string and tie a reef knot. Separate the strands of string so that they bush out from the knot.

12 Fray the edges of the calico, then scrunch it up in the middle, using clips to hold the shape. Heat the glue gun and apply glue to the scrunched folds. Press the calico onto the center of the box lid.

13 Remove the clips from the calico and apply more hot glue. Press the bark bundle on top of the calico.

14 If you are using the box as a card index, write labelling on some spare "bark" paper and glue this to one end of the box.

Right: The range of options for country-look boxes is numerous, and brown wrapping paper provides an excellent base for further decoration. Gilt wax rubbed into the paper will give a golden hue. Then add decorations from nature such as dried fruits or leaves. Natural strings such as garden twine or raffia make excellent ties if the box is to be a gift.

CUTLERY BOX

This project fuses the clarity of high-tech design with the weirdness of surrealist sculpture – and it provides an ideal place to keep your cutlery at the same time! The stylish boxes look especially good in a modern kitchen, where chrome and stainless steel keep the lines crisp and the surfaces reflective. Make separate boxes for knives, forks, and spoons and say goodbye to rummaging in the kitchen drawer in search of the right implement.

YOU WILL NEED

- small silver-plated knife, fork and spoon, polished
- 3 metal boxes, with lids
- felt-tipped marker pen
- coarse-grade sandpaper
- metal file
- metal-bonding compound
- craft knife (or other fine instrument)

1 Bend the knife to a right angle halfway along the handle. It should bend easily, but, if not, do it over the edge of a table.

2 Place the knife on one of the boxes and mark its position. Roughen the contact point on the knife with sandpaper. Carefully rub the part of the lid that will make contact with the knife handle with a file.

3 Mix the metal bonding, following the manufacturer's instructions. Apply the bonding to the roughened area on the lid. The knife is fixed only at this point, so the bond needs to be strong.

4 Press the knife handle into position on the bonding. Use a fine instrument, such as a craft knife, to remove any excess bonding. Repeat all the steps for the fork and spoon.

MATERIALS

The materials used in this book all require different treatments. Your basic tool kit should include a glue gun, a staple gun and a cordless drill, as well as the usual hammer, screwdriver, tape measure, pliers, and scissors. A craft knife with a set of blades is essential. If you feel that you would get enough use out of it, buy an eyelet punch, or, as an alternative you can buy cheaper eyelet packs with disposable tools.

Spray adhesive is useful since it can be lightly sprayed to provide a low-tack fixing that allows the separation of the two surfaces at a later time – this makes it ideal for holding stencils in place during painting. Sandpaper and steel wool are graded according to coarseness and they can be used to fine tune wooden surfaces. Steel wool is particularly good for giving painted wood an aged look.

You need paints that suit the surface you are treating, so follow the instructions given for each project. For brilliant color on paper or cardboard, try premixed watercolor or acrylic paint. These come in droppered bottles and can be used to tint any waterbased medium, such as latex paint, white glue, wallpaper paste, or acrylic varnish.

Gather together a collection of buttons, braids, string, and general bits and pieces. This way you will avoid the hassle of having to start from scratch each time inspiration for a container strikes!

Chicken wire (1); staple gun (2); string (3); high-density foam (4); acrylic paints (5); shellac (6); aerosol (spray) paint (7); fabric paints (8); brilliant watercolors (pre-mixed) (9); white glue (10); craft knife/utility knife (11); paintbrushes (12); center punch (13); scissors (14); long-nosed pliers (15); awl (16); wire cutters (17); hammer (18); brass eyelets (19); glue gun (20); tape measure (21); upholstery tacks (22); eyelet punch (23); steel wool (24); sandpaper (25); Dutch metal leaf (26)

25
24
26
23
4
21
22
10
20
11
13
12
14
15
19
18
17
16

TECHNIQUES

Everybody has their own way of doing things, but sometimes the "proper" way is a complete revelation and actually quite helpful. It has often developed out of everyone else's hit-and-miss attempts and, although it might seem like stating the obvious, here are a few basic guidelines to point you in the right direction.

Using an electric drill

Using an electric drill for the first time can seem quite daunting, so read the manufacturer's instructions carefully beforehand. A cordless drill makes life easier as you do not have to work near a plug. Drill bits and wallplugs are number-coded to make sure that you always use the same size. To drill a hole to the correct depth, mark the length of the wallplug on the drill bit with masking tape before you make the hole.

1 To drill into plaster, first mark the position in pencil. Then cover it with a small square of transparent tape. If you press the tape onto the back of your hand before applying it to the wall, it becomes less tacky and is unlikely to lift any paint after removal. Drill through the tape using a masonry bit.

2 Tap the wallplug gently with a hammer until it is level with the wall. The plug should fit snugly for a strong fixing.

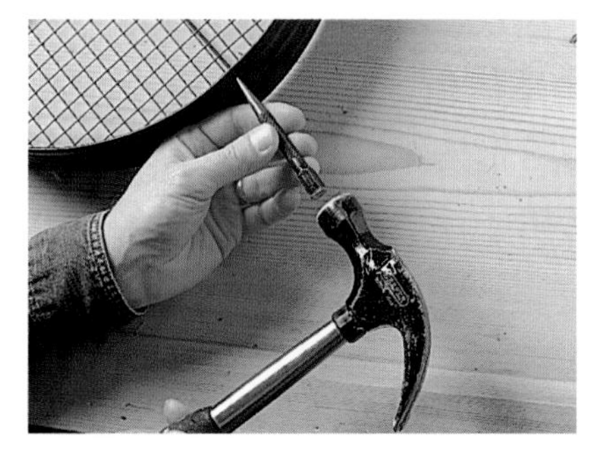

3 To drill into metal, use a standard metal drill bit. Tap a dent in the metal using a center punch and a hammer. If you do this, the drill will not skid off when you start it up.

4 If you are using a variable-speed drill, start it off slowly and do not operate it at full speed. Metal heats up under this sort of friction, so either clamp it or wear protective gloves if the metal is thicker than a tin can. Use wet and dry sandpaper to get rid of any burr that has formed around the hole.

Sanding and sealing

1 Using a medium-grade sandpaper, remove the top layer of wood and round off any sharp corners. Then rub all over the surface with fine-grade sandpaper to give a really smooth finish.

2 Apply a coat of shellac button polish. This is the best sealant for bare wood, and it dries to a yellowish sheen. Clean the paintbrush right away with methylated alcohol.

3 Dampen a soft cloth with turpentine and wipe the surface to clear it of any residual dust. Apply another coat of polish.

Using antiquing varnish

Antiquing varnish can be either homemade or bought in a selection of shades that imitate old wood finishes. Antique pine is the lightest and is best for general use. Follow the manufacturer's instructions for drying times and paintbrush cleaning. The varnish is a light toffee-brown and adds a yellowish tone when dry.

1 To make your own antiquing varnish, place a tablespoon of clear matte water-based varnish onto a plate and add a ¾in squeeze of raw umber acrylic or watercolor paint. (If you are using polyurethane varnish, you should use an oil paint instead.) Mix well using a paintbrush.

2 Pick up some of the varnish with a soft cloth and rub it into the surface, working it well into the cracks and crevices to cause a darker buildup of color. You can test the shade on an inconspicuous part of the box first. For a darker shade, use burnt umber.

3 Rub the box with fine sandpaper or steel wool, then dampen a cloth with turpentine and wipe the surface gently. Allow to dry.

4 Using steel wool, rub the places where natural wear and tear would be most obvious, to reveal the wood and paint underneath.

Tinting varnish

If you find the overall color of your container too harsh, you can use varnish to mellow it. A good way to get the exact shade you want is to tint acrylic varnish with waterbased paint. You can add yellow to warm up browns, reds, and oranges, or use blue and green for the opposite effect. Tinted varnish creates subtle color changes and adds a protective coating at the same time.

Place a tablespoon of waterbased varnish onto a plate and add a small amount of golden-brown waterbased paint. Mix together with a paintbrush, and test the shade on an inconspicuous part of the box. Adjust the shade by adding more paint or white glue. Then brush it over the surface. Apply a second coat, if necessary.

Apply varnish with a good quality clean paintbrush, working with parallel strokes in the same direction. Apply enough varnish to cover the wood completely, but not too much because it may run and spoil the finish.

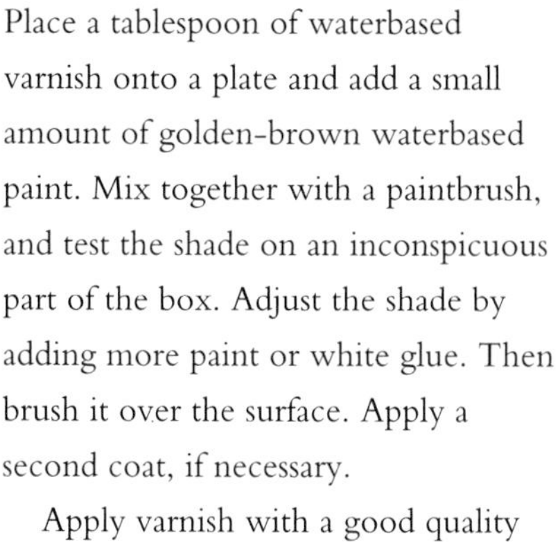

TEMPLATES

1 2 3

4 5

6 7 8

9 0

INDEX

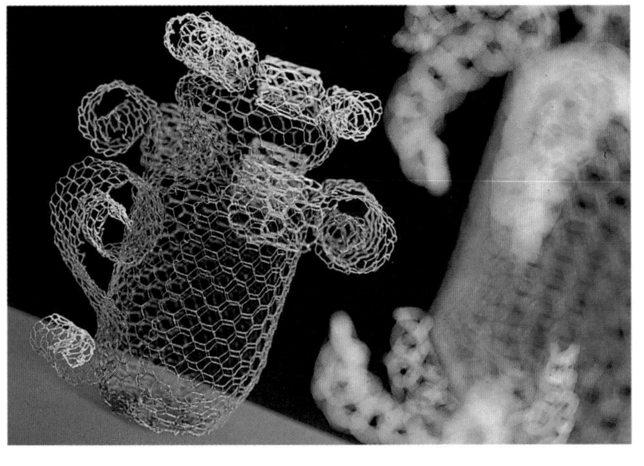

The authors and publishers would like to thank After Noah for generously supplying all peculiar wire baskets and metal boxes used in this book:
After Noah, 121 Upper Street
Islington, London N1 1QP,
United Kingdom